Energy Education: Goals and Practices

By Rodney F. Allen

Library of Congress Catalog Card Number: 79-93115
ISBN 0-87367-139-2

Bloomington, Indiana

This fastback is sponsored by the Triad North Carolina Chapter of Phi Delta Kappa, which made a generous financial contribution toward publication costs.

TABLE OF CONTENTS

Energy and Crisis

Waiting in line at gas stations during the winter of 1973-74 as a result of an oil embargo was a new experience for most Americans. On the night of 11 November 1975 New York City and the Northeast were in darkness because of an electrical blackout. In several summers of the 1970s consumers were faced with a series of brownouts. In the winter of 1977 natural gas and oil supplies were insufficient to meet the demand. In 1979 political waves in Iran and the Middle East brought odd/even gas selling days to the states on both coasts of the U.S.

Americans learned the terms *oil crisis, energy crisis,* and *peak load* under duress. Energy was at the forefront on the national consciousness. *Renewable* and *nonrenewable, fission, fusion* and similar terms became common parlance in legislative halls, in lunchrooms, and in the home. Our national vocabulary was enriched while our pocketbooks and energy reserves were depleted. Conservation took on personal meaning.

The local press carried articles on the latest scheme to turn Thoreau's Maine woods into alcohol or the Arizona desert into a mammoth solar furnace. The national media flooded us with data on oil imports, balance of payments, nuclear plant construction, and the latest word on windmills as an energy source. Doomsday forecasts alternated with energy conservation advice in the nation's press.

Americans received a contemporary history lesson. Abundant energy had revolutionized the ways that we lived with, thought about, valued, and interacted with one another. Life-styles, industrial and agricultural production patterns, and human patterns of intercourse all had been shaped by the abundance of cheap energy. Government regulation of oil prices fostered the belief that this cheap energy would continue to be available indefinitely. Our historical assumptions about national progress and personal prosperity were solidly based on the mass utilization of inexpensive energy. From water, wind, wood, and mule, we leaped to coal, oil, and natural gas—with nuclear power more than a hope on the horizon. Before the "crisis" struck in the mid-1970s, our personal and national energy appetite had become insatiable. Figures 1 and 2 summarize the growth in energy consumption. Figure 2 is particularly disturbing, because it shows how our consumption of energy continued to grow, in some years exponentially.

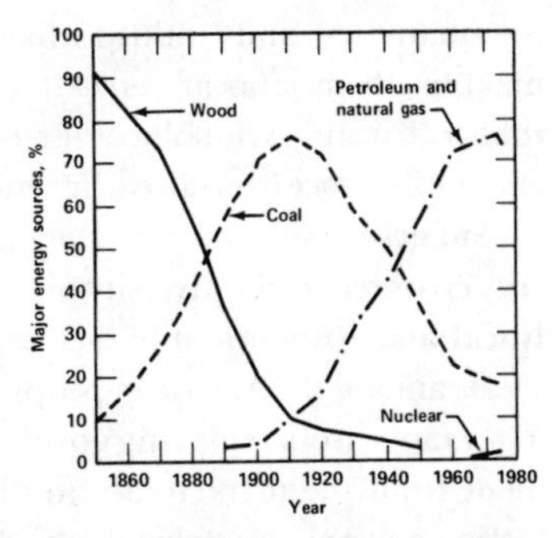

Fig. 1. Historical U.S. energy consumption patterns

Source: U.S. Energy Research and Development Administration, *Creating Energy Choices for the Future*, Washington, DC, 1975, p. 2.

During the winter of 1973, when the energy crisis affected the personal life-styles of most Americans, the response of many was to search for a villain. The oil producers served as convenient scapegoats for some; environmentalists who blocked the development of new energy sources caught the ire of others. Today, with soaring electric bills and increasing gasoline costs, most Americans are less convinced there is only one villain.

The media offered Americans a plethora of explanations for the energy crisis. Commentators with an orientation toward economics saw the fundamental cause in the marketplace, where true energy costs were not translated into consumer prices. They argued that such factors as the scarcity of usable energy, damage to the ecology, net-energy concepts, and incentives for new energy development and for conservation measures were not consistently reflected in pricing mechanisms. Commentators from the political right blamed too much governmental interference and regulation. Those from the left pointed to financiers and industrialists, making them appear as callous robber barons, symbols of greed and indifference from an earlier era.

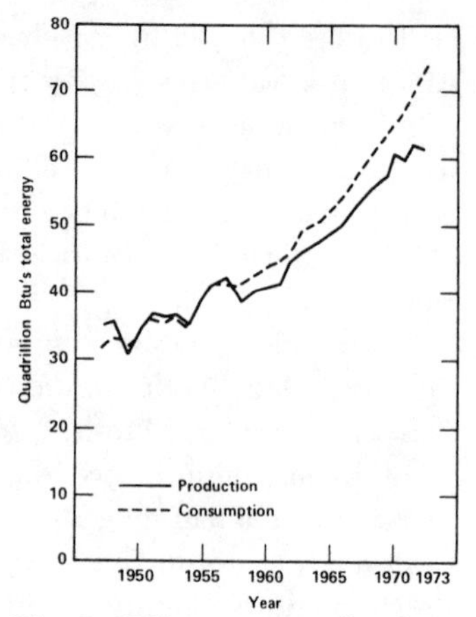

Fig. 2. U.S. energy production and consumption, 1947-1973

Source: *Exploring Energy Choices*, Ford Foundation, Washington, D.C, 1974, p. 2.

Suggested solutions to the energy crisis reflect the diversity of political and economic interests and ideals among the American people. They range from removing government from the marketplace to directing government to break up the vertical integration of energy companies; from government ownership of the sources and production of energy to increasing government incentives for research, exploration, and capital formation.

Today more Americans realize that the energy crisis is real. It is not just one more in a long series of doomsday forecasts, nor is it another crisis hyped by the media and then forgotten as the new crises arrive to capture the headlines. We now realize that there is no such thing as a free lunch in energy. We now realize that in using energy we can never break even—we always lose energy in converting it to work for us.

Today Americans are less disposed to look for a villain and more disposed to search for solutions. We are all, in our own way, the cause in a very complex international problem. There is no one solution to our problem. There is no one technological advance, no one political decision that will turn our situation around. It will take personal sacrifice, consistent and morally sensitive public policies, and energy research and development to achieve a solution.

The time for decision is now. In the past decade we have watched the lethargic and piecemeal response of governments to energy supply and demand problems. There have been a few gestures and a lot of pronouncements. Certainly, we can do better.

Our schools have been hit by the rising costs of energy and the uncertainty of oil deliveries. Across the nation, school personnel are conducting energy audits on school buildings. But what are educators doing about energy education? Yes, more science teachers are devoting additional instructional time to energy concepts and principles. Social studies teachers, as part of citizenship education, are presenting energy policy issues in their classrooms. Many home economics teachers have turned to the energy implications of consumerism and home management. Yet, there has been *no clear call* among educators for instruction about energy in the general education of today's students. Can we build an appropriate *educational* response to today's energy crisis and to our students' energy future?

Why Study About Energy?

All that we can predict with certainty is that the central issue of the twenty-first century, as it is this one, will be the struggle to assert truly human values and to achieve their ascendency in a mass, technological society.

—John Goodlad

The Greek root of the word *crisis* means "decision." The ancient Chinese wrote the word *crisis* with two characters: one meant "danger," and the other meant "opportunity." The problems and issues inherent in the energy crisis reflect those ancient meanings—danger, decision, opportunity. There is a danger that requires thoughtful decisions among competing choices. But the crisis offers us, as individuals and as a people, the opportunity to work together toward a solution.

Two national administrations have sought to develop an energy agenda for the U.S. Both administrations have forecast the implications of the energy crisis and have noted the critical importance of learning about energy and of learning to conserve energy. During the Nixon-Ford administration, the Federal Energy Administration (FEA) worked with groups in all sectors of the economy and at all levels of government on energy allocation plans and energy conservation programs. By 1975 the Energy Research and Development Administration (ERDA) proposed a national energy plan to meet the growing shortfall between domestic energy demand and domestic energy production. The goals set in the ERDA plan involved preserving national security, maintaining environmental quality, cooperating on energy policies with the world community, and preparing for economic transitions due to changing energy realities. At the same time, the ERDA plan was working to fulfill the life-style aspirations of Americans—especially the less affluent. With these goals in mind, the ERDA plan placed a high priority upon factors of energy supply. Money was to be spent enhancing the recovery of oil and gas. Coal was to be used more as a primary fuel, and nuclear energy (including breeder fission reactors) was to be greatly expanded. In addition, the ERDA plan called for increased funding of research on innovative energy technologies.

A second priority was reducing demand for energy, that is, energy conservation. As a technical strategy, ERDA approached energy conservation by stressing innovative technology to increase the efficiency of energy use. Research and development efforts would yield new processes that business and industry would apply with their own capital. However, for the most part, federal energy conservation policy was based upon voluntary reduction in the per capita energy demand in the U.S. The ERDA plan had little to say about how such voluntary reductions were to occur, but it was clear that the energy problem was in large measure due to this per capita increase in energy demand. ERDA instituted some training programs for teachers, using summer workshops and conferences. The FEA funded several small curriculum development efforts directed toward school children and their teachers.

In 1977 the Carter Administration announced its energy plan, calling the energy crisis "the moral equivalent of war." Over the long term, the plan sought renewable and "inexhaustible" sources of energy to sustain economic growth and a high quality of life for all Americans. The shorter term goals focused on reducing oil imports to lessen the vulnerability of the nation. The immediate goals related directly to conservation: reduce the annual growth in energy demand to below 2%; reduce gasoline consumption by 10%; use solar energy in more than 2.5 million homes; and bring 90% of American homes and all new buildings up to minimum energy efficiency standards. This emphasis upon conservation was a continuation of the Nixon-Ford emphasis. It was clear to those in energy policy-making positions that the only short-term strategy to solve the national energy dilemma was conservation, and its corollary, improved energy efficiency.

While headlines and public discussions of the national energy plans most often focused upon technology and taxation, a knowledgeable citizenry ready and willing to conserve was the cornerstone of our energy conservation policy. President Carter addressed the matter of will, discipline, and understanding, using strong moral terms:

> The ultimate question is whether this society is willing to exercise the internal discipline to select and pursue a coherent set of policies well in advance of a threatened disaster. Western democracies have demonstrated such discipline in the past in reacting to immediate, palpable threats to survival, as in time of war. But they have had less success in harnessing their human and material resources to deal with less visible and immediate threats to their political and economic systems. When dangers appear incrementally and the day of reckoning seems far in the future, democratic political leaders have been reluctant to take decisive and perhaps unpopular action. But such action will be required to meet the energy crisis. If the nation continues to drift, it will do so in an increasingly perilous sea.
>
> *Executive Office of the President, 1977*

But the drift continues. Electricity consumption goes up each year. The demand for gasoline countinues, despite rising prices and lines at the pumps. The amount of energy used in food production in the United States increases each year. Building design has not been altered appreciably by prices for energy or the prospect of harsh energy futures. Public surveys show limited knowledge about energy among the citizenry. Surveys in early 1979 showed that less than 50% of the adult population knew that the U.S. imports oil; the rest thought that domestic production did and will satisfy our demand. Those same adults are no more knowledgeable about the connectedness of energy facts: linking electricity for a light bulb with oil and natural gas; linking Mexican strawberries and California lettuce with the availability of diesel fuel; or linking dripping hot water faucets with electric or gas bills. This lack of knowledge is compounded by a lack of confidence in government policy and a suspicion of the so-called "special interests" in the energy production and distribution systems.

Energy Conservation and Education

Energy conservation is the quickest way to bring about rapid reduction in our use of energy. Of course, energy conservation via more efficient use of energy is a long-term goal involving innovative design and capital investment. But increasing energy awareness, increasing the knowledge base of energy consumers, and increasing the incentives for saving energy are means that can be applied today.

But what are the incentives for motivating people to conserve energy in a free society? There are three routes to bringing about such conservation. First, the government (at various levels) has considerable coercive power. Using legislation and administrative rules, the government can manipulate the self-interest of energy consumers through windfall profits taxes, strict building codes, insulation tax credits, and odd/even days at the gas pumps, for example.

Second, social norms can be used to reward and punish. In the past, social norms linked prestige with large gas guzzlers. By imposing new norms, we can learn to value energy-conserving vehicles and look askance at those who persist in using oversized cars driven at outrageous speeds. Norms define social expectations and impose social sanctions. Social attitudes about energy consumption and new conceptions of the "good life" can mold our behavior: living close to work and walking, riding mass transportation, creating bicycle paths, and adjusting the course of urban development (e.g., zoning) with energy realities in mind. Both of these routes obviously involve influencing the personal decisions of others by "external" force. While laws, rules, and social norms are instructive as well as coercive, a free society must be wary about the extent to which it relies on coercion.

A third route to encourage energy conservation is voluntarism, involving personal decisions based on knowledge of the energy crisis and on a commitment to regulate one's behavior in behalf of both self-interest and the community. Discussions of energy conservation policy based upon voluntarism have been few—reflecting, perhaps, a diminished confidence in the people's willingness to rally around a vital issue out of a sense of responsibility to the community. Yet, voluntarism and rational personal decision making are driving forces in a free society and they must not be overlooked.

Of course, there is a bottom line to be reckoned with when human attempts at energy conservation fail. As the fossil fuels are diminished, individual and societal decisions will be overruled by the irrevocable natural laws of our ecosystems. When supplies of oil and natural gas run out, the element of human choice is removed. Societies and individuals can only respond by seeking other alternatives. Conservation of fossil fuels is no longer an issue. Likewise, as one depletes the forest near the village and has to walk farther and farther each day for firewood, the price of wood goes up. To burn dung for warmth or for cooking food is a choice—but a harsh one. No free society and no rational person wishes to be regulated by such harsh choices imposed by the final limits of the ecosystem. In fact, this curtailment of choice in matters of public policy and personal decision making could be the death knell of a free society.

Both the Nixon-Ford and the Carter Administrations have sponsored legislation and exerted political leadership to influence societal norms regarding energy use. They have also proposed education as a means of modifying personal decisions regarding energy consumption. It is clear, however, that the energy crisis issues are moral, political, technical, and economic. The technical and economic tasks ahead are demanding, but the moral and political issues will require tremendous educational efforts.

Given the reality of the energy crisis, it is reasonable to conclude that:

1. Energy conservation is the most effective short-term response to our energy realities and an absolute necessity over the longer term.

2. Energy conservation requires knowledge of energy facts and the personal commitment to act on those facts in order to save energy—both of which can be implemented with a systematic energy education program.

3. Citizens will have to participate in the formulation of new social norms to cope with future energy realities. This will require hard knowledge, new skills, and a commitment to change attitudes about energy use for the common good.

In the past our society has looked to education to help deal with many national crises. We have sufficient data now to anticipate a future of harsh energy shortages. The goal of energy literacy is one that educators must accept and society will expect.

Justifications for Energy Education

Often energy education is defined as the teaching of basic concepts and facts about energy phenomena. Also, energy education is defined as promoting energy conservation wherein the goal is not only learning facts and concepts but taking appropriate action as well. In both cases, the arguments for such programs in the schools seem to be based upon one or more of the following justifications.

Educated Citizenry Justification: A basic premise of American democracy is that citizens with adequate skills and knowledge participate in making the decisions that affect them. Solutions to energy problems require a knowledge of, and participation in, the political process. Traditionally, education has served to equip citizens with the knowledge and skills needed to make such decisions and to take action.

Economic Justification: Energy education is needed so that citizens may make rational choices in their personal lives about public policy questions.

"Doing Good" Justification: Energy education offers the opportunity to create a more just and humane society. It can provide the re-educative process that will lead to a more environmentally sound lifestyle and a more equitable distribution of goods and services in the national and world community.

Problem-Solving Justification: Solving widespread energy and environmental problems can be undertaken by a well-educated citizenry. Since energy is so directly related to the conditions that sustain life and give it meaning, educational programs are needed to provide the techniques for solving the energy crisis.

"Take Care of Self" Justification: Energy education can teach people how to conserve energy and to make decisions to optimize their personal resources and well-being. Such learning will build personal responsibility and will help to protect personal effects.

Educated Person Justification: Society is beset with technological problems about resource use that require educated and inquiring minds. Energy education should be part of general education now and in the twenty-first century. A sound knowledge of energy concepts and issues should be expected of all educated people.

Career Education Justification: Many of today's students will find careers in energy industries (e.g., solar installation, energy audits, and energy-efficient construction) or in industries that depend on crucial energy decisions. Schools have a role in preparing students for these and other careers.

Stewardship Justification: Energy education, like environmental education, offers an opportunity for persons to learn a "common good" ethic, where the well-being of all people (including those yet unborn) is taken seriously in resource use and in the allocation of goods and services. If everything is connected to everything else, we need to learn those relationships and to use them. The welfare of all and responsibility to the posterity of our nation are central themes in this argument.

Apocalyptic Justification: We must learn the wise use of energy resources since there is such a limited supply of nonrenewable energy resources. National survival and the well-being of the world community depend upon effective and equitable use of energy to ward off resource depletion on the one hand and environmental disaster on the other. We need people who can stand at Armageddon and do battle for our survival.

Whether or not one concurs with all of the justifications offered here for energy education programs in the schools, learning about energy is essential for the social education of all children whose lives will extend well into the twenty-first century. In responding to dwindling energy supplies, our technological advances may be spectacular. Legislation may force us into immediate conservation efforts. Pricing mechanisms may cause life-style changes. But in the long run, it is the knowledge, abilities, and personal commitments of the people in a free society that will permit successful responses to the energy crisis.

Goals for Energy Education

Energy education is a national and a personal priority. Appropriate instructional activities can have a direct and immediate impact upon energy conservation and patterns of energy consumption. Those responsible for planning instruction will need to attend systematically to the goals of energy education for specific student audiences.

Vocational-technical educators are training students for the growing job market in energy conservation technical jobs and for energy conservation technology as applied to traditional occupational fields. Goals in such programs include:

1. Learning how to perform energy audits of existing buildings
2. Learning how to install solar devices and cost-effective insulation
3. Learning how to increase energy efficiency in equipment (e.g., tuning automobiles, adjusting air conditioners, adjusting thermostats)
4. Learning how to manage utility consumption in facility operations (e.g., school cafeterias, hotels, factories, etc.)
5. Learning how to position and landscape buildings for effective use of environmental conditions and solar energy installations

Adult basic educators are helping students develop basic survival skills in an urban-industrial society. Energy competencies are a new addition to the list of basic skills traditionally taught. These competencies stress practical techniques with immediate application:

1. Reading and adjusting a thermostat for heating and cooling
2. Selecting appropriate lighting fixtures and lighting intensity

3. Changing air filters

4. Monitoring energy consumption and utility bills

5. Learning household management techniques to save energy dollars

6. Changing washers on leaky hot water faucets

7. Caulking or weatherstripping household windows and doors

8. Selecting automobiles and appropriate driving speeds

9. Insulating hot water lines and tanks

The general goal is to help persons make and carry out energy and financial decisions based upon enlightened self-interest. That is, they should know when and how to conserve energy in order to achieve their personal life goals and at the same time contribute to energy conservation in their communities.

For the general education of children, youth, and adults, the goals of energy instruction are better stated in language that describes the broad range of competencies necessary for persons to function as citizens and as rational consumers. The following list of such goals has been adapted from Fowler (1976):

1. The student-citizen will understand the science and technology of energy and its pervasive role in the universe, in living systems, and in human systems (e.g., national, global, political, economic, and social systems).

2. The student-citizen will be able to make informed and equitable judgments on energy options as they arise, being able and willing to participate in the political decision-making process.

3. The student-citizen will be able to make personal life-style commitments that are consistent with energy realities and are morally responsible.

4. The student-citizen will be aware of and prepared to participate in opportunities for setting energy policy and encourage energy conservation at the personal, local, state, and national levels.

The general education of student-citizens should include a multidisciplinary knowledge base. They need to know what energy is, the kinds and sources of energy, and the nonrenewable character of fossil fuel resources for urban-industrial societies. Concepts from the natural and physical sciences and from the social sciences must be mastered to comprehend the complex role of energy in natural and social systems. It is not adequate to understand simply one facet, such as the energy flows of the biosphere, without looking at energy flows in one's own community and its economic system. For example, to understand electricity pricing requires knowledge of concepts related to the conversion of energy and energy loss; concepts related to the production of electricity with various technologies (including nuclear and solar); economic concepts of capital investment, marketing, pricing, and distribution systems; and government regulation. In addition, utility rate structures involve public policy issues, such as environmental quality trade-offs and distribution of electricity to low-income customers faced with inflationary utility bills.

An often overlooked, but equally important, aspect of the energy knowledge base is understanding the interrelationships between daily behavior and energy use. Switching on a light, adding a new home in the neighborhood, and driving 10 miles to play tennis have an impact upon energy use, resource allocation, and capital formation. Letting a hot water faucet drip will increase the rate at which the household gas or electric meter turns. While these relationships may not seem as profound as global energy forecasts, economic cost-benefit analyses, or the science of solar generation of electricity, the relationships are fundamental to coping with daily life, building a base for energy conservation, and using energy resources more efficiently.

A conceptual scheme for the knowledge base of energy education was developed at Florida State University. Stated as main ideas, each component leads to inquiry and reflective lessons appropriate at various grade levels. For example, third-grade children may examine conflict over human rights, interests, and obligations through a unit of study exploring the choices in a family budget during a very cold winter. Older students might examine the same idea through the analysis of electric rates in their area. Adults in their community education program could develop the idea through a philosophical inquiry relating issues of human rights to the national energy plans and Congressional debates on energy policy issues.

The main ideas are grouped into three strands: 1) the universe of energy, 2) living systems and energy, and 3) social systems and energy.

The universe of energy and living systems and energy strands outline basic scientific ideas about energy and energy flow models of which people are a part. The social systems and energy strands set forth fundamental concepts in social systems, including governmental, economic, and moral systems, the understanding of which is vital in making decisions affecting the production, distribution, and consumption of energy resources. The three strands described below are extensively rewritten from a scheme developed by the John Muir Institute for Environmental Studies (1974).

Strand I. The Universe of Energy

1. Energy is the ability to do work.
2. Energy exists in many different forms, including light energy, electrical energy, chemical energy, mechanical energy, and heat energy.
3. Changes in the motion or position of matter occur only when energy is exerted.
4. Almost all of the energy available on earth comes directly or indirectly from the sun.
5. The earth is an open system that constantly receives solar energy and gives off heat energy.
6. Life can exist on earth only because of the constant and steady arrival of solar energy and the equally steady loss of heat energy into outer space.
7. Machines and living organisms change energy from one form to another.
8. Energy can be changed in form, but it can never be created or destroyed.
9. Different forms of energy are able to do different amounts of work.
10. Kinetic energy refers to any form of energy that is actively doing work.
11. Potential energy refers to any form of energy that is inactive or stored. All matter contains potential energy.
12. Sources of energy for future use include organic matter, nuclear materials, and solar energy.

Strand II. Living Systems and Energy

1. All living organisms require energy for such functions as movement, responsiveness, growth, reproduction, and metabolism.

2. Green plants (producers) are the only form of life that can capture the energy available in solar radiation. They do it by means of a chemical reaction called photosynthesis.

3. Organisms (such as humans) that cannot capture solar energy obtain energy from green plants either directly or indirectly. These nongreen organisms are consumers.

4. Organisms that break down dead animals and plants into molecules and atoms are decomposers. Decomposers are usually bacteria and fungi.

5. Feeding relationships between producers, consumers, and decomposers form patterns called food chains or food webs that describe the paths by which energy is transferred from one organism to another.

6. The overall pattern formed by the movement of energy from producers to consumers is a complex food web called an energy pyramid.

7. As energy flows through a living system, it imposes order and organization on that system.

8. Under certain conditions, energy stored in the tissue of dead organisms may become a fossil fuel.

Strand III. Social Systems and Energy

1. All people must consume energy to stay alive.

2. People transform and manipulate energy sources to satisfy their needs and wants.

3. People use energy to improve their environmental conditions, to power machines, and to maintain their culture.

4. People are among the very few organisms that use large quantities of energy resources.

5. People have increased their consumption of energy resources throughout history.

6. People living in technological cultures have greatly increased their consumption of energy resources in the last few hundred years.

7. Sources of energy have changed as new types have been found and as old sources have been depleted or found to be less desirable.

8. The major sources of energy have changed from renewable ones, such as plants and animals, to depletable ones, such as coal, oil, and natural gas.

9. People use energy to create and sustain special ecosystems, such as cities, recreation areas, and agricultural areas.

10. People have used energy resources to increase agricultural yields and thus increase the amount of food energy available to them.

11. People are beginning to look toward energy resources that are nondepletable.

12. All societies have wants greater than their resources are able to fulfill, creating the condition of scarcity. Economic systems, governmental systems, and moral systems are used to give direction in allocating scarce resources, including energy resources.

13. Energy consumers have interests, obligations, rights, and ideals that govern their personal and collective consumption of energy resources.

14. Energy producers have interests, obligations, rights, and ideals that govern their production and distribution of energy resources.

15. Social systems, including government, the economy, and societal networks, have interests, obligations, ideals, and rules affecting the production, distribution, and consumption of energy resources.

16. Individuals, groups, and the society at large face conflicts in self-interests, obligations, rights, and ideals as they make choices or rules affecting the production, distribution, and consumption of energy resources.

17. Energy conservation deals with increasing the efficiency of energy use and decreasing the amount of energy used.

18. Social systems that regulate energy supplies and use are important components in energy conservation.

19. People as energy consumers and decision makers are individually and collectively responsible for energy conservation.

Mastery of these main ideas provides a sound knowledge base for the general education of student-citizens. To these statements of knowledge, instructional programs must add practical competencies in knowing how to conserve energy and to involve student-citizens in confronting the value conflicts inherent in energy policy questions.

Moral Development Goals in Energy Education

When President Carter announced his energy plan, he was hard-pressed to come up with a "common good" concept that would touch the moral fiber of the citizens of this nation. The best he could do was to borrow William James's phrase, "the moral equivalent of war." In America, we tend to operate on a distributive ethic (Who gets what and why?) rather than an ethic based on the common good. Environmental issues, and now energy issues, are basically "common good" issues—issues that our national traditions have not prepared us to handle. Our core civic values are liberty and equality, both individual in nature. We assess the health of our society by how much equality there is or how much liberty we enjoy.

Our conception of justice assumes a community of citizens engaged in a common life. Persons in the community have a sense of dignity and a set of roles to play, each with honor. People are bound together by a system of shared values and regard for one another. But somehow in our growth we have turned these ideals into a rampant individualism that means promoting one's self-interests to the limit of one's power. We define ourselves by our personal rights and opportunities, not by our debts and legacies. Hence, the President had difficulty coming up with a symbol of public philosophy in his appeal to American citizens to curb self-interest for the good of all.

The directive calls for us to reestablish civic idealism. Edward Schwartz, president of the Institute for the Study of Civic Values, has urged educators to replace individualism, privatism, hierarchy, and obedience, with cooperation, equality, and participation. That is, we need to restore a sense of community and a covenanted patriotism, using the principles of the Declaration of Independence and the Constitution in light of current realities. We need to revive the ideals in these documents with regard for what is "good" for all. The ideals in these documents contain the common body of sentiments for legitimatizing energy policies that make sense for the future. Alexis deTocqueville, traveling in America during the 1830s, grasped the essence of the

American character when he wrote in *Democracy in America*:

> The American moralists do not profess that men ought to sacrifice themselves for their fellow creatures because it is noble to make such sacrifices, but they boldly aver that such sacrifices are as necessary to him who imposes them upon himself as to him for whose sake they are made. . . . They do not deny that every man may follow his own interest, but they endeavor to prove that it is in the interest of every man to be virtuous. . . . *The principle of self-interest rightly understood* [author's italics] produces no great acts of self-sacrifice, but it suggests daily small acts of self-denial. . . . It disciplines a number of persons in habits of regularity, temperance, moderation, foresight, self-command.

Thus, energy education may be perceived as part of the more general process of social education. In this context, the goals for energy education need not be stated only in terms of ways to save energy or to learn energy concepts. Rather, the goals are intended to promote the personal/social fulfillment of persons as moral agents in their local, national, and world communities. The study of energy issues is not an end in itself, but rather is a route to more important life goals. In dealing with energy issues we can sort out our value priorities and assert truly human values in a mass, technological society. In studying about the energy crisis we have the opportunity to reaffirm our confidence in our collective ability to face issues and meet the challenges. As Kenneth Clark observed:

> Of course, civilization requires a modicum of material prosperity—enough to provide a little leisure. But far more, it requires *confidence*—confidence in the society in which one lives, belief in its philosophy, belief in its laws, and confidence in one's own mental powers. . . .

A Model for Energy Education

There has been a tendency in science and social studies to present data and abstract concepts far removed from the experience of the learners, and this has been true for energy education programs as well. Also, because of the crisis situation, there has been a tendency to issue warnings and threats about what will happen if we don't learn to conserve energy. These are not viable *educational* methods.

Energy education requires a model based on the students' reality, one that does not indoctrinate and one that sees information and concepts as instrumental learning. Such a model should take into account:

Students' personal awareness of what is going on around them and how they are responding inside (perceptions, beliefs, feelings, values, and actions).

Students' personal meaning of the reality they see surrounding them in society and within themselves (beliefs, feelings, values, judgments, and explanations).

Students' personal evaluation of what is happening and of what they are feeling, valuing, judging, and doing, given their values and goals.

Students' ability to make decisions, after evaluating alternatives, as to the appropriate way to act.

Students' ability to employ skills for deciding how to act, given the judgments they have made (e.g., developing a strategy for acting, an action plan).

Students' motivation to do something about themselves (their way of feeling, thinking, valuing, acting) and about their natural/social environment.

Students can learn facts about natural gas, the causes of the energy crisis, and methods of conserving energy, but this model for energy education goes beyond cognitive enlightenment. It provides a process to use hard knowledge from the sciences and from experience to develop personal meaning, social values, and action plans. Both the process and the content are crucial to the model.

The chart below summarizes the process of the energy education model. A stimulus or springboard activity is used to obtain a high level

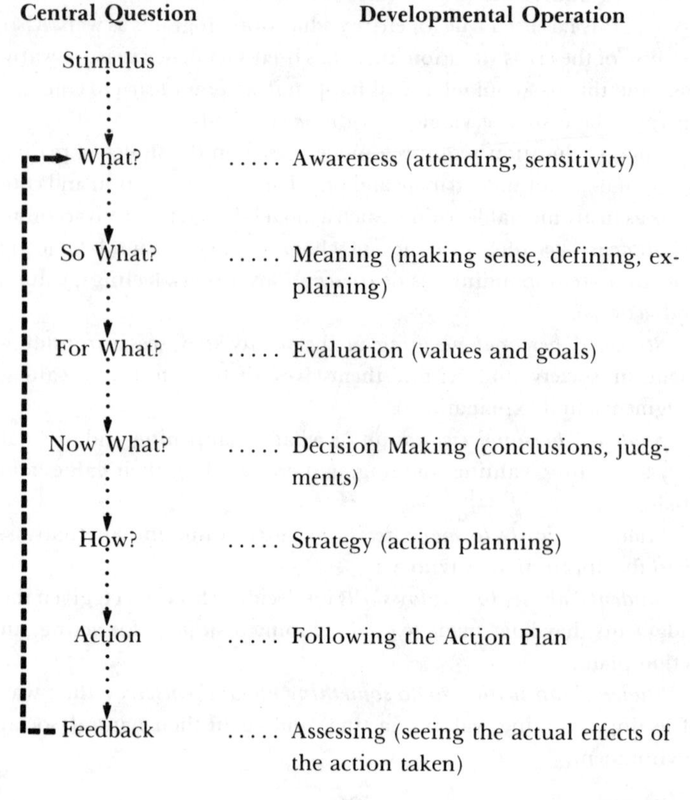

of interest, then a series of simple questions allow the teacher to explore the multidimensional facets of energy issues under consideration. Each question is carefully linked to a cognitive or affective developmental operation.

How the Model Works

The opening of any lesson begins with a perplexing stimulus. The stimulus may be an ice cube, a picture of a dripping faucet, a song, a poem, a thermostat, or a burning candle. These stimuli evoke curiosity.

The teacher then raises *What?* questions. The *What?* questions focus student attention upon their own perceptions of what is going on in a certain setting, raising personal awareness.

The teacher then presents *So what?* questions. Students reflect on the meaning of what they perceive or feel. Why is the ice cube melting? Did you feel sorry to see the ice vanish? What happens to heat when a door is left open? So what if poor people must spend more of their pay for electricity than wealthy folks?

These questions are followed by *For what?* questions. Students are asked to evaluate what they have experienced, perceived, or felt. What is worthwhile about what we saw? Given what you value, or a goal you want to attain, what is wrong with hot water faucets dripping? Basing your judgment on the information you collected, would it be better to insulate the floor or the ceiling to conserve heat? Is energy loss desirable? Is it fair to charge small consumers more for electricity?

Evaluative questions are then followed by *Now what?* questions. What ought to be? What should we do? Students are asked to make interpretations and judgments about how they should act, think, or feel. What do we want to do now? What action should we take? What goals should we pursue? Are our goals fair to others? Are our goals good for us? Why?

Deciding what to do is one step in the process, but students also need to decide how to act to change their thinking, feelings, or values. *How?* activities are used to assist students in building an action plan to produce desired results. How can we get the school district to buy energy-efficient buses? How do we install insulation? How much does it cost? How can we raise the necessary funds?

Action simply provides the opportunity for students to do something they have decided upon: paint a picture, turn off lights, conduct a home energy audit, caulk school windows, rethink their feelings about poor people's energy costs, or work for more energy-efficient school buses.

Feedback is an assessment of the results of action taken. Students act in order to attain desired effects. Now they should reflect upon the actual effects of their behavior. Have you altered your feelings about the poor? How have you changed? Are you making progress on the school bus issue? What effect has caulking the windows had? What went right for you? What went wrong? What can we do differently to be more effective?

Many skills are developed and used in this process. Communication skills, inquiry skills, reasoning skills, and creativity have a part to play. Discussion skills (acting together with others, sharing, etc.) are critical. Students need to know how to share their ideas and feelings. They need to learn how to pose inquiry questions, shape hypotheses, and collect data. They need to be able to give reasons to justify conclusions. They need to learn how to question another person's judgments. And, of course, students need the opportunity to develop social skills so that they can speak coherently before a group, write letters to officials, and work sensitively and effectively with others in a group.

Model Lessons

Five model lessons for energy education follow. The first three were developed at Florida State University by George O. Dawson, David E. LaHart, Marvin Patterson, and Rodney Allen. The last two were developed by the Lawrence Hall of Science staff for the U.S. Department of Energy (see appendix).

These lessons were selected as brief examples of the model described above. Longer lessons for secondary school students may be found in the curriculum materials listed in the appendix, especially those produced by the National Science Teachers Association *Project for an Energy Enriched Curriculum.*

Lesson 1. A Dripping Hot Water Faucet

Idea: A dripping hot water faucet wastes both the fresh water and the energy used to heat the water.

Materials: A hot water faucet in the classroom or laboratory that may be allowed to drip for a few minutes

A clock or stop watch

Different kinds of measuring devices (e.g., a measuring cup and a metric calibrated cylinder)

Action: Have students observe the dripping hot water faucet. Ask them to identify a problem. (Students should share ideas.) (*What?*)

Activity 1. Try to determine how much hot water is being lost. How would you measure water loss using the materials provided by your teacher? (*What?*) (*How?*) Discuss your plan with another student, then do it.

Note to Teacher: Get students to design and carry out an investigation to determine water loss. They will need to consider time and volume measures. Let them first attempt to solve the problem and then raise the need for their establishing standard measuring devices. If some students collect a cup of water in 10 minutes and others collect 300 ml in 12 minutes, they will have a problem in comparing data. Have the students establish a standard, such as the amount of water that will be wasted in 10 minutes, or how long does it take to collect 250 ml of water.

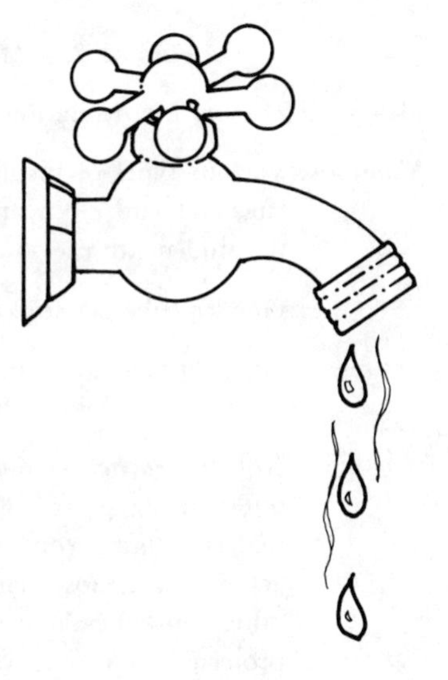

Activity 2. Compare your information on the amount of hot water lost with another student's information. What problems do you have in making comparisons? Discuss with your teacher the problem of comparing information with the rest of the class.

Going Further: 1. Have you seen dripping hot water faucets before? Where? What did you do? Why did you do that? (*So what?*)

2. Develop an action plan to correct drippy hot water faucets at home or in school. You might begin by doing a "Drippy Hot Water Faucet Survey" to see how many dripping faucets you have in school or at home. (*How?*)

3. Pretend that you are a guest at Sally's home. You see dripping hot water faucets in her kitchen and in her bathroom. You do not want to hurt her feelings. What should you do? (*How?*)

Lesson 2. A Melting Ice Cube

Idea: How to use insulation to conserve heat. (*What?*)

Materials: Various types of insulation material such as paper, corrugated board, cloth, and Styrofoam should be available for the students to use.

One ice cube per student or pair of students.

Action: Have the students plan how they might keep their ice cube from melting longer than anyone else's. (*So what?*)

Note to Teacher: Give students time to plan. Let them carry out their suggestions. Do not tell them to use the insulation materials, but permit them to if they ask. At the end of the activity, have those that were most successful share with the others how they kept the ice from melting and why they protected the ice the way they did. (*For what?*)

Going Further: To explore the social consequences of this phenomena, ask students to relate what they learned that will affect decisions when buying the following items. "What will you think about when you and your parents are buying:

a container to take coffee to a ball game?"

an ice chest to go camping or fishing?"

an oven for your kitchen?"

a freezer or refrigerator for your home?" (*Now what?*)

Lesson 3. Color and Heat

Idea: Does the color of the roof make a difference in the temperature of your house? (*What?*)

Materials: Model of a house with removable roof, one side painted white, the other side black.

Thermometer

Heat source (sun or sun lamp)

Action: Activity 1. Place the thermometer on a hook in the middle of the house. Place the roof on the house with the black side facing out. Place the house in the sun or under the heat lamp. Wait five minutes and read the temperature of the air inside the house.

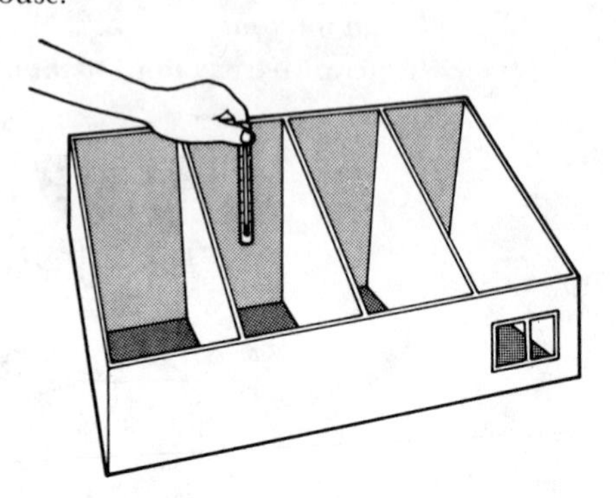

Activity 2. Place the roof on the house with the white side facing out. Place the house in the sun or under the heat lamp. Wait five minutes and read the temperature of the air inside the house.

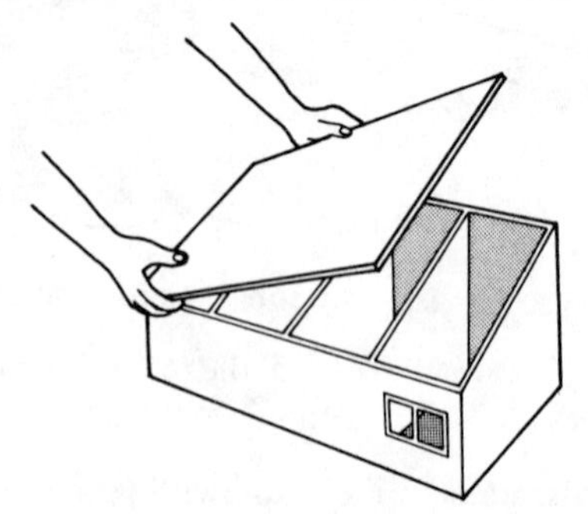

Activity 3. 1. While you are waiting, answer the following question: Do you think that the temperature in the house with the white roof will be higher than, lower than, or the same as the tempera-

ture in the house with the black roof? Why? (*So what?*)

2. Make two graphs while you are waiting. First make a graph showing what you think the temperature difference will be. Then using the thermometer readings, make another graph showing the actual readings.

3. What statements can you now make about the value of having a white roof if you live in Florida? (*For what?*)

Going Further: 1. Work in small groups to list ways to make a home cool in a warm climate or warm in a cold climate. Share your lists with one another. (*How?*)

2. With your teacher's and your parents' help, arrange to talk to home builders, insulation installers, nursery owners, and other community workers about your list. Ask them to help you add to your list. Get them to help you to figure out how much each item on the list would cost for your home. (*How?*)

3. Work with your principal or custodian to figure out ways to make your classroom cooler in warm weather and warmer in winter. (*How?*)

Lesson 4. Tire Pressure and Energy

Idea: Will your bicycle coast twice as far if your tires have twice the pressure? (*What?*)

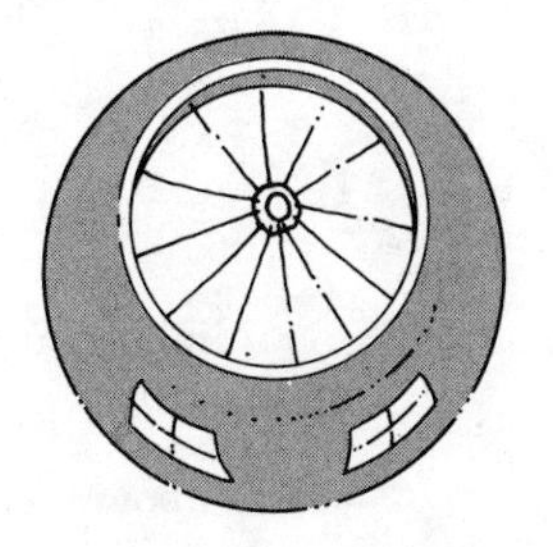

Materials: 2 similiar bicycles
1 tire pressure gauge

Action: Activity 1. Inflate one bicycle's tires to normal pressure and the other to half that amount.

Activity 2. Have two students of similar weight ride side by side at the same speed. When they reach a selected line marked on the ground, they should coast until they come to a stop. Compare how far each goes.

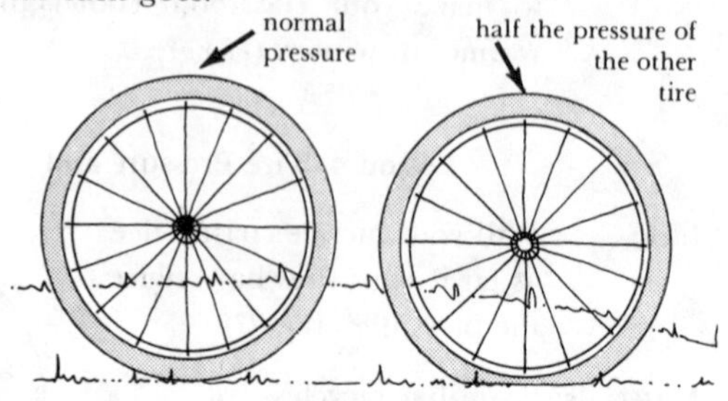

Is it important to check the tire pressure on your bicycle? (*So what?*)

How about your family car? (*So what?*)

Lesson 5. Drafty Houses and Energy

Idea: Is your house drafty? (*What?*)

Materials: Pencil
Scotch tape
Plastic food wrap

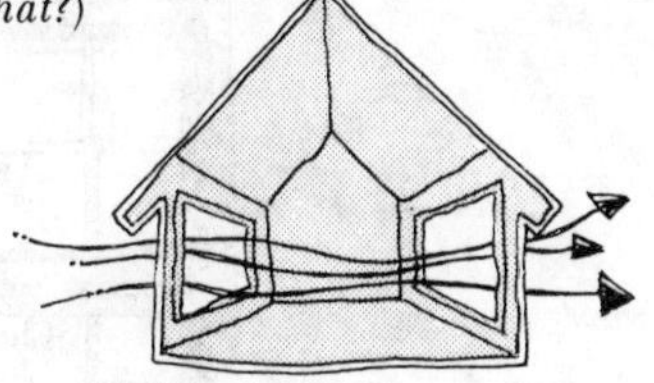

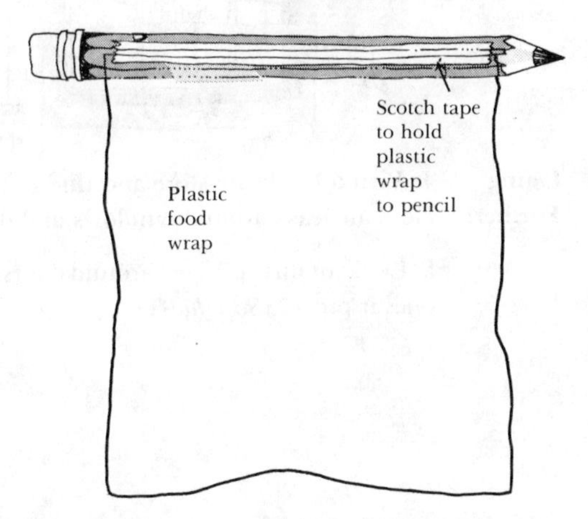

Action: Activity 1. Make a draftometer by following these instructions: Cut a 12 cm by 25 cm strip of plastic food wrap. Tape it to the pencil. Blow the plastic gently and see how freely it responds to air movement.

NOTE: Forced-air furnace must be off to use draftometer!

Activity 2. Test your home for air leakage by holding the draftometer near the edges of windows and doors.

Test your fireplace with the damper open and closed. Why are drafts energy wasters?

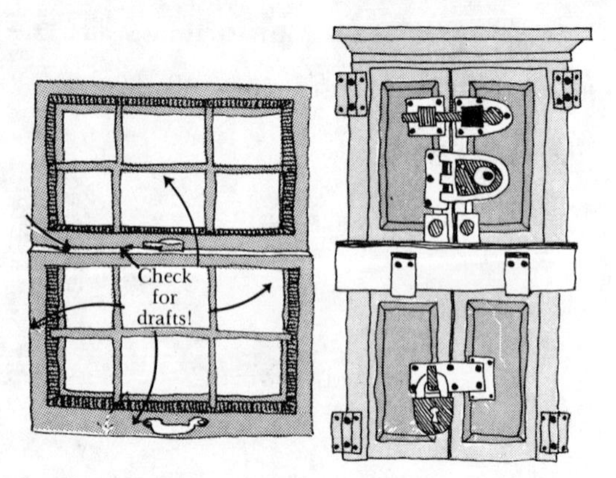

Going Further:

1. Visit a hardware store and find out what is available to close air leaks around windows and doors.

2. Look for dirt collected around doors and windows. What does it prove? (*So what?*)

Introducing Energy Issues into the Curriculum

Schools wanting to include energy education in their curriculum will discover a plethora of materials. Commercial interests ranging from oil companies to public utility companies are sending materials to teachers promoting the energy producers' views. Publishers are turning out booklets and multimedia modules in attractive packages. Federal agencies are sending out free and inexpensive materials reflecting the views of the particular agency. State and local educational systems are drafting energy education curricula for teachers to use in their classrooms. Organizations such as the National Science Teachers Association and the Biological Sciences Curriculum Study have written and tested a number of instructional units. These efforts offer educators an array of energy information and instructional materials.

There are three possible ways to build energy education into the curriculum: a) adding a new unit, b) adding a new course, or c) infusing the curriculum with information on energy. It will be difficult to add a new required course to the already crowded middle or high school curriculum. Elective courses are a possibility, especially in the vocational/technical curriculum, but they would not reach a very high percentage of the students. Adding a unit has been a popular response, but it is difficult to see much lasting impact of another three- to five-week unit. Also, it is getting difficult to tack on another unit in most courses. However, curriculum infusion has promise as a way to incorporate energy education. U.S. history teachers could deal with the history of energy use when studying the industrial revolution, the westward movement, or the urbanization of America. Geography teachers could highlight energy resources and commerce in many of their traditional units. Science teachers could treat the energy flow of natural systems in biology courses, the physics of solar energy in physical science courses, or the processes of nuclear fission and fusion in chemistry courses. Elementary school teachers in their science and social studies units could readily deal with conservation studies, energy use studies, and such special topics as Transportation and Energy, Community Helpers and Energy Use, and Energy Audits for Our Classroom. This curriculum infusion strategy permits teachers to insert energy concepts, principles, and issues where they are appropriate into ongoing units of study. Teachers concerned about "basic skills" will find that they can incorporate energy topics in the students' study of reading, writing, and mathematics.

Creative teachers in all subjects will have no difficulty introducing energy content into their courses. While there is a great variety of free and inexpensive materials available, teachers must take care to see that sponsored materials are free of bias and indoctrination. No matter how worthy the intentions of the producers of energy education materials, they should be rejected if they set out to impose fixed beliefs or to gain acceptance of certain policies that sacrifice the long-term development of the learners for the short-term inculcation of "good energy habits or beliefs."* The test for teachers reviewing instructional materials should be, "Does this teaching material preserve the integrity of learners and foster their development into *rational* participants in a free society?" Also, teachers, whose primary charge is serving students, should have the responsibility of determining what is taught in energy education and how it is taught. Teachers and school administrators need to reserve their right to accept or reject educational materials and opportunities (e.g., field trips, student programs, etc.) offered to them by outside groups and agencies.

---

*For a report of research on materials produced by corporations, see Sheila Harty's *Hucksters in the Classroom: A Review of Industry Propaganda in Schools*, available from the Center for Study of Responsive Law, P.O. Box 19367, Washington, DC 20036. $10.

Conclusion

In proposing a national energy plan in the spring of 1977, President Carter observed that

> . . . meeting the nation's energy goals should be a great national cooperative effort that enlists the imagination and talents of all Americans.

The same cooperative spirit, enlisting the talents and imagination of governmental agencies, school officials, community leaders, and classroom teachers, applies to the goals of energy education. No single agency or organization ought to impose educational goals or curriculum materials upon teachers and students. Rather, a cooperative effort can provide students in various communities with the best possible understanding of energy concepts, principles, and issues and give them the skills and know-how to deal with harsh energy realities. That is both the hope and the promise of energy education in our schools.

Appendix
Sources of Practical Curriculum Materials for Energy Education

Deciding How to Live on Spaceship Earth (1973). McDougal, Littell Publishing Company, P.O. Box 1667, Evanston, IL 60204. Price, $3.50; 135 pages.

An ethics case book designed for secondary school students, it offers decision-making situations involving energy, natural resources, and environmental issues. A teacher's guide is available from the publisher.

The Economics of the Energy Problem (1975). Joint Council on Economic Education, 1212 Avenue of the Americas, New York, NY 10036. Price, $1.50; 20 pages.

One issue of the Economic Topic Series, this publication offers educators an overview of the energy crisis in terms of supply, demand, cost, and capital requirements, with extensive teaching suggestions for high school classes. Many brief case studies are included.

Edison Experiments. Thomas Alva Edison Foundation, Cambridge Office Plaza, Suite 43, 18280 West Ten Mile Road, Southfield, MI 48075. Prices vary.

The Edison Foundation publishes a series of energy booklets for students. Some are devoted to historical themes and others focus on hands-on electrical experiments. Titles include: *How to Build Five Useful Electrical Devices* (1971); *Thomas Alva Edison's Associate, Lewis Howard Latimer, A Black Inventor* (1973); and *Electrical and Chemical Experiments from Edison* (1973).

The Energy Crisis: An Introduction, by Mary Jo Leavitt and Harry Leavitt (1977). United Graphics, Inc., 1428 Harvard, Seattle, WA 98122. Price available from the publisher; 52 pages.

This packet of materials consists of duplicating masters that are self-contained and self-directed. The guidebook offers adequate background for teachers.

Energy-Environment Source Book, by John M. Fowler (1979), and *Energy-Environment Materials Guide,* by Katheryn Mervine and

Rebecca Cawley (1975). National Science Teachers Association, 1742 Connecticut Ave., N.W., Washington, DC 20009. Price $6, pre-paid; *Source Book,* 279 pages; *Materials Guide,* 60 pages.

The Energy-Environment Source Book offers teachers basic energy concepts and facts and deals with the many issues and problems in energy development. *The Energy-Environment Materials Guide* offers educators an annotated bibliography of the most authoritative and accessible energy literature, a graded list of student readings, and a guide to films and other nontext materials.

Energy Management Center. Pasco County Schools, P.O. Box 190, Port Richey, FL 33568. Tom Baird, project director. Units and kit plans available at cost.

Funded under ESEA Title IV-C and the Pasco County (Florida) School Board, this project developed instructional programs on energy and energy conservation for fourth- and ninth-grade students. Project units are complemented by visits to the Energy Management Center for demonstrations and hands-on experiments and by portable kits for research and study on the school campus. The units and kits are readily adapted to other regional school settings in North America.

Energy and Society: Investigations in Decision-Making (1978). Biological Sciences Curriculum Study. Hubbard Scientific, P.O. Box 104, Northbrook, IL 60062. Contact publisher for prices.

This package of materials is designed for a nine-week instructional unit for high school and adult education students. It presents data on current energy realities, offers alternative consequences of energy decisions, provides basic energy concepts, and guides students through research on an energy problem of their selection. The unit comes with an instructor's manual, a student handbook, a set of projection slides, a silent 8mm filmloop, a card set, and an energy management game.

Fact Sheets on Alternative Energy Sources (1977). Technical Information Center, U.S. Department of Energy, Box 62, Oak Ridge, TN 37830. Free upon request.

Prepared by the National Science Teachers Association under contract with the U.S. Department of Energy, the set includes 19 fact sheets, each four to eight pages long, on a variety of energy topics: Energy Conservation, Alternative Energy Sources: A Bibliography, A Glossary of Terms, Breeder Reactors, Geothermal Energy, Nuclear Fusion, Wind Power, Solar Heating and Cooling, etc. The fact sheets are useful background reading for educators and for above-average high school readers.

99 Ways to a Simple Lifestyle. Reports Department, Center for Science in the Public Interest, 1757 S St., N.W., Washington, DC 20009. Price, $5.50; 324 pages.

This volume contains practical ideas and information for energy conservation. The major areas covered include heating and cooling, energy conservation in the home, solid waste, transportation, food, and recreation. Well-illustrated.

People and Energy Newsletter. Center for Science in the Public Interest, 1757 S St., N.W., Washington, DC 20009. Subscriptions: $7.50 individuals, $15 libraries.

This newsletter is published monthly. It contains research findings related to energy issues, announcements for recent publications, and forthcoming government hearings and conferences.

Energy and Education Newsletter. National Science Teachers Association, 1742 Connecticut Ave., N.W., Washington, DC 20009. Available free upon request.

Published under contract with the U.S. Department of Energy, this quarterly newsletter contains information about recent developments and announcements of future events of concern to educators and others interested in energy education.

Energy and Man's Environment (1973). Education Research System, Inc., 2121 Fifth Ave., Seattle, WA 98121. $2.95 plus postage and handling.

This interdisciplinary guide contains a rationale for and sample activities in energy education, K to 12. It provides springboards for educators' use in developing their own curriculum materials.

Energy and Order, by Mark Terry and Paul Witt (1976). Friends of the Earth, 529 Commercial St., San Francisco, CA 94111. Price, $3.

This is a five-week unit on basic energy concepts stressing the social impact of energy use and conservation.

The Energy Challenge (1976). Federal Energy Administration, Box 14306, Dayton, OH 45414. Single copies free.

Distributed under contract with the Federal Energy Administration (now the U.S. Department of Energy), the materials in this packet are ditto masters, each offering information on energy consumption, conservation, production, and supply.

Energy Conservation Training Institute Manual. The Conservation Foundation, 1717 Massachusetts Ave., N.W., Washington, DC 20036. Available from the Foundation.

Under contract with the Federal Energy Administration (now the U.S. Department of Energy), the Conservation Foundation produced this guide to train local citizens in public policy options for energy conservation efforts. The manual contains 20 papers, each covering one aspect of the issues involved in developing energy conservation strategies (such as transportation, taxation, land use, utilities, and environmental impacts).

Project for an Energy-Enriched Curriculum (PEEC). National Science Teachers Association, 1742 Connecticut Ave., N.W., Washington, DC 20009. Materials available free upon request from the Technical Information Center, U.S. Department of Energy, Box 62, Oak Ridge, TN 37830.

Under the direction of John M. Fowler, curriculum writers and teachers are developing units covering the first through the twelfth grades. Fifteen units are currently available, each offering learners energy information, energy concepts, and inquiry opportunities related to energy issues and problems. Sample unit topics include: *The Energy We Use* (Grades 1 and 2); *Community Workers and the Energy They Use* (Grades 2 and 3); *Bringing Energy to the People: Ghana and the U.S.* (Grades 6 and 7); *How a Bill Becomes a Law to Conserve Energy* (Grades 9 through 11); *U.S. Energy Policy: Which Direction?* (Grades 11 and 12); *Mathematics in Energy* (Grades 7 and 8); and *Networks: How Energy Links People, Goods, and Services* (Grades 4 and 5). Each unit is designed to be infused into regular school course offerings, especially in social studies, science, and mathematics.

Science Activities in Energy. Technical Information Center, U.S.

Department of Energy, Box 62, Oak Ridge, TN 37830. Free upon request.

Prepared by the Lawrence Hall of Science, Berkeley, California, under a 1977 contract with the U.S. Department of Energy. This series contains six titles; each offers an array of hands-on energy experiments: *Chemical Energy* (26 pages); *Conservation* (28 pages); *Electrical Energy* (30 pages); *Solar Energy* (23 pages); *Solar Energy II* (30 pages); and *Wind Energy* (30 pages).

Organizations Active in Energy Issues and Education

American Association for the Advancement of Science
1515 Massachusetts Ave., N.W.
Washington, DC 20005

American Conservation Association, Inc.
30 Rockefeller Plaza
New York, NY 10020

American Petroleum Institute
1801 K St., N.W.
Washington, DC 20006

Consumer Federation of America
Suite 901, 1012 12th St., N.W.
Washington, DC 20005

Cooperative Extension Service
County (office usually at county seat)
State (office usually on campus of land-grant university)

Energy Conservation Corps
c/o The Bolton Institute
1835 K St., N.W.
Washington, DC 20006

Environmental Protection Agency
Public Information Center (PM 215)
Room 2106
Washington, DC 20460

Hatheway Environmental Education Institute
Massachusetts Audubon Society
Lincoln, MA 01773

League of Women Voters
1730 M St., N.W.
Washington, DC 20036

National Recreation and Park Association
1601 North Kent St.
Arlington, VA 22209

National Wildlife Federation
1412 16th St., N.W.
Washington, DC 20036

Scientists' Institute for Public Information
30 East 68th St.
New York, NY 10021

State Energy Office
c/o The Governor's Office
(state capital)

U.S. Department of Energy
Office of Communications and Public Affairs
Washington, DC 20461

Bibliography

Clark, Kenneth. *Civilization: A Personal View*. New York: Harper and Row, 1970.

Executive Office of the President, *The National Energy Plan*. Washington, D.C.: Government Printing Office, 1977.

Fowler, John M. "Energy, Education and the 'Wolf Criers.'" *Social Education* (April 1976): 251-257.

Glass, H. Bentley. *Science and Ethical Values*. Chapel Hill, North Carolina: University of North Carolina Press, 1965, pp. 82-84, 89-101.

John Muir Institute for Environmental Studies, Berkeley, California. Schema produced under the terms of a 1973-1974 grant from the U.S. Office of Environmental Education, Grant #OEG-0-73-5450, David B. Sutton, director.

Litke, Robert F. "What's Wrong with Closing Minds?" In *Values Education*, edited by John R. Meyer, et al. Waterloo, Ontario: Wilfrid Laurier University Press, 1975.

U.S. Energy Research and Development Administration, *Creating Energy for the Future*. Washington, D.C.: Government Printing Office, 1975.